The French Way of War:
Eight Strategy Lessons for Today's Conflicts

Michael Shurkin

The French Way of War:
Eight Strategy Lessons for Today's Conflicts

Pax Americana Press

ISBN 979-8-234-03140-2

For Rafael, Noam, and Saadia

Acknowledgements

This little work would not exist had I not had the pleasure of meeting with and talking to French officers over the years, men who, delighted by my interest, pointed me in right directions. If these officers are as competent as they are intelligent—and I believe they are—France is well served. I particularly wish to thank General (ret.) Charles Beaudouin, Colonel (ret.) Fabrice Clee, Colonel (ret.) Betrand Darras, General Armel Dirou, General Pascal Facon, Colonel (ret.) Michel Goya, General (ret.) Philippe Gueguen, Colonel Rémy Hémez, General (ret.) Guy Hubin, Colonel Jean Michelin, General Philippe de Montenon, and General Hervé Pierre for their friendship and guidance.

I also wish to thank Alan and Lili Kalish Gersch for their love and support.

Introduction

If I had to sum up the French school of military strategy in only a few words, I would fall back on *Ne pas subir*—which translates awkwardly as "Do not be subjected to" or "do not submit." This was the catch phrase of Marshal Jean de Lattre de Tassigny (1889-1952), known to the French as de Lattre or sometimes even "Le Roi Jean," or King Jean.

De Lattre is said to be one of 20th century France's best fighting commanders. He fought at Verdun in World War I and in the Battle of France in 1940. Though he served under Vichy, he rallied to Charles de Gaulle in 1943 and was a senior commander of the First French Army that landed in southern France in 1944 and fought its way into Germany in 1945. After the war, he rose to the rank of the senior commander of the French military, and in 1951 became France's senior civilian and military commander in Indochina before dying of cancer in 1952. He was posthumously promoted "Marshal of France."

De Lattre inspired subsequent generations of French commanders and had a profound influence on one of his subordinates in 1944-1945 and Indochina, General André Beaufre, who is the greatest of modern France's military strategists. François Géré, in his 2017 study of contemporary French military strategy, insists that it all started with de Lattre.[1]

---

[1] François Géré, *La pensée stratégique française contemporaine* (Economica, 2017). For more on de Lattre's influence on Beaufre and the two men's relationship, see General Hervé Pierre, *Le Général Beaufre: Père de la stratégie française* (Perrin, 2025).

I cite de Lattre because his catchphrase neatly sums up an approach to strategy shared by Beaufre and his intellectual successors as well as two major forbears, Marshal Ferdinand Foch—the mastermind of the allied victory over Germany in 1918—and Admiral Raoul Castex, one of the greatest naval theorists of the 20th century alongside the United Kingdom's Julian Corbett. It is an approach to strategy that, while being distinctly French, nonetheless is universal in application and useful for its clarity. The following work offers a distillation of that approach.

Some readers might look askance at the idea of turning to the French for strategy lessons given the 20th century French military's many debacles. I argue on the contrary that defeat does more to inspire reflection than victory. One cannot rest on one's laurels if there are no laurels upon which to rest. There's no place for complacency among men who felt they could not be complacent. They learned hard lessons. In contrast, I have always found American military thinking to be diminished by complacency and, above all, a lack of self-doubt. That leads to an absence of reflection. Shallowness. Perhaps worse, American military thinking often bears the marks of hubris. Hubris, like gout, can be thought of as a rich man's disease. It is an affliction associated with over-abundance. French military thinkers suffer from many problems, but hubris is not one of them. How could it be?

Definition of Strategy

Before proceeding, it is necessary to define strategy. Here I turn to Beaufre. Strategy is the "art of the dialectic of wills that employ force to resolve their conflict."[2] In this dialectic, the "decision" each

---

[2] General André Beaufre, *Introduction à la stratégie* (Pluriel, 2012), 34.

2

side seeks to impose is psychological, not material. It amounts to convincing the adversary that engaging in or pursuing a struggle is useless. Beaufre argued that strategy was a "duel of wills" that produced "the opposition of two symmetrical games." Each side sought to "strike the decisive point of the other" by means of efforts intended to "frighten, paralyze, and surprise." The goal, really, was primarily "psychological," meaning it was not in fact necessary to destroy the enemy but rather to induce in the adversary a state that allowed one to impose one's will upon it.

Beaufre argued that strategy had "two distinct and essential elements." The first was the "choice of a decisive point that one wants to strike." This, he explained, was a function of one's assessment of the enemy's "vulnerabilities." The second was a "choice" of a "preparatory maneuver" that would make it possible to strike the decisive point. But here was the key thing: Given that each side was attempting to do the same thing, victory came to the side that stopped the adversary's maneuver and "conducts his own to its objective."

Approach

This work is structured around eight concepts that together make French military strategy distinct. No strategist worth reading claims to possess a recipe for success or what today one might call a "hack" that guaranteed victory. What the French strategists referred to here offered instead was a mental framework for thinking about how to win this duel of wills. The intent here is not to provide a scholarly discussion of the thinking behind each theme, or who contributed what to the concept. This is neither a history of military strategy nor an academic study. Rather, the aim is to distill useful concepts from a corpus of writing and teach them.

The eight concepts, one chapter each, are as follows:

- Liberty of Action
- Initiative
- Maneuver
- *Effet majeur*
- The Primacy of Will
- Indirect Strategy
- Total Strategy
- Servitudes

Liberty of Action and Initiative in a way are self-explanatory. It is a question of understanding their importance in the context of strategy. Maneuver, *Effet majeur*, and Will similarly have particular meanings when it comes to strategy, most notably with respect to the critical task of maintaining and expanding one's Liberty of Action. Indirect Strategy is a concept meant to help one think about how best to advance to one's objectives, or, to put it differently, to reach the enemy's decisive point. Total Strategy refers to the need to subordinate military strategy to a total strategy of which military strategy is but one complementary component. Lastly, Servitudes refers to a concept charted by Castex. It has to do with the relationship between military strategies and other kinds of strategy that together should comprise a total or Grand Strategy.

Liberty of Action

Liberty of Action is at the heart of strategy. If strategy is a struggle of wills, that struggle boils down to a struggle for Liberty of Action, with each seeking to conserve it while denying it to the adversary. As long as one possesses Liberty of Action, one can maneuver. One can take the initiative. One can do *something* according to one's will, ideally something that increases one's ability to continue to act according to one's will. Without Liberty of Action, one is subject to the will of one's adversary. One is acted upon. In war, this is tantamount to defeat.

In classical, conventional warfare of the sort evoked by Marshal Ferdinand Foch in his 1903 *Principles of War*, which he grounded in careful study of the campaigns of Napoleon Bonaparte, the principal way of protecting one's Liberty of Action is by deploying Security, often in the form of a force deployed in front, the sides, or perhaps behind one's main force. One objective of deploying such a force is to hide from the adversary one's main force if the smaller detachment bumps into the enemy. The enemy must determine if it has encountered one's main force, which wins some time. The smaller force, moreover, diverts at least some of the enemy's attention and resources. The enemy might even mistake the smaller force for the main force and engage it accordingly. More importantly, the smaller force can report back to the main force, granting the main force the ability to act according to its will, rather than act reflexively or otherwise be compelled to take any specific course of action. In other words, the smaller force grants the main force Liberty of

Action: The main force can choose to engage, or withdraw, or move to the side. It does not *have to* do anything it does not wish.

An analogy for Security is extending one's arm when entering a dark room or perhaps feeling first with one's foot. The idea is to sense an obstacle before bumping bodily into it. Once sensed, one can react thoughtfully, perhaps by avoiding the obstacle, or ignoring it. Otherwise, one might trip or crash into something. In that case, one is committed to a response rather than able still to respond as one chooses.

Of course, Security implies dividing up one's force; it requires Economy of Force, meaning knowing how to "apportion one's means rationally between protecting against the adverse preparatory maneuver, one's own preparatory maneuver, and the decisive action." [3] This is contrary to the notion of Concentration of Force, of being able to gather all of one's resources at one place and time. But Security gives one the Liberty to do that at the right time and place, to decide when and where to concentrate one's means.

In modern conflicts, preserving one's own Liberty of Action and diminishing that of the enemy is more complicated. One set of constraints might be classified in terms of Servitudes, to be discussed below. They boil down to the idea that military commanders are not free to pursue their primary military objectives without having to consider other priorities that are foreign to pure military strategy. A major one, of course, is *policy*. War, as Clausewitz infamously asserted, is politics by other means, but the reality, as even Clausewitz understood, went deeper than that. War had to serve political objectives and therefore how one fought and with what means ultimately depended on policy considerations outside the normal realm of military strategy.

---

[3] Beaufre, *Introduction à la stratégie*, 52.

In the 20th century, policy would constrain military strategy far more than it did in Clausewitz's day or even Foch's. This became particularly true with the advent of nuclear weapons, when total war (in the sense of the kind of all-out effort to destroy the enemy that one saw with the two World Wars) became impossible, at least among nuclear powers. Nuclear weapons, however, did not preclude the possibility of limited wars. Arguably they made limited wars more likely, for only acts of aggression calculated to be below the threshold of triggering a nuclear response were possible if one wanted to gain an advantage over an adversary. Limited wars implied careful political calculations: Unable to go all out to destroy the enemy, one has to identify limited objectives and measure out the limited means thought sufficient to achieve those objectives. (It is difficult at the moment not to think of the Iran war, which is taking place as I write: The U.S. could have opted to go all out to destroy Iran but has opted instead for a limited war, which requires identifying limited objectives and the limited means understood to be sufficient to achieve them. Has Washington done this?)

In the post-Hiroshima era, the enemy retains the means to limit one's Liberty of Action, even when demonstrably weaker. Beaufre described these as the external maneuver, for it took place outside the theater of the war, and more on the international stage. The idea was to use international forums, law (lawfare), and propaganda to tie one's adversary's hands, rather like the Lilliputians tied down Gulliver. The North Vietnamese, for example, deployed propaganda on the international stage and recruited celebrities like Jane Fonda as a means of pressuring the United States government to curtail its strategic bombing campaigns. The result may have been more effective at blunting the U.S. Air Force than North Vietnam's air defenses.

Today there is no better example than Palestinian armed groups and their allies' constant efforts to use international law and international forums like the United Nations to limit Israel's ability to wage war. Their aim is to nullify as much as possible Israel's military advantages and limit its Liberty of Action by making military maneuvers diplomatically, economically, and socially costly. One might not be able to defeat the Israeli Army in battle, but one can harass Israelis at Eurovision, or Jewish students on American university campuses. Something as seemingly benign as street demonstrations can lead to boycotts of spare parts for fighter jets, which could have a far greater effect on Israeli Air Force operations than attempts to counter it with air defense missiles. One might lack the naval power to deter the Israeli navy and force it to abandon its Gaza blockade, but one can place pressure on Israel to lift the blockade by deploying flotillas of civilians in the hope of grabbing media attention and stoking public outrage. Thus, something as seemingly innocent as advocating for banning Israel from participating in international sports tournaments is neither trivial nor peripheral to the Arab-Israeli conflict but rather a classic example of external maneuver.

Similarly, Russia has been trying to limit Ukraine's combat effectiveness by promoting in Europe and the United States propaganda intended to undermine political resolve to keep funneling arms and money to Kiev or encourage the United States to throttle back intelligence sharing. There is nothing innocent about all those social media posts hostile to supporting Ukraine and its war effort; preventing or slowing arms deliveries is perhaps even more effective than interdicting weapons on the battlefield, for it involves minimal risk and spares Russian forces from having to do the job themselves.

Often countries impose upon themselves limitations on their Liberty of Action, usually in light of policy considerations or attention paid to the Servitudes, to be discussed below. Parties to a conflict have a sense of what they can and cannot do lest they lose international or domestic support, for example, or risk escalation to unacceptable levels.

For example, France in 1914 just prior to the outbreak of war, kept the bulk of its forces well away from the German border to avoid appearing to be the aggressor if war began. Even though French doctrine at the time was profoundly offensive in nature, the French government believed it could not make the first move and accepted the idea that it had to allow Germany to attack first. Their concern was keeping Britain on side. This meant that France voluntarily ruled out any pre-emptive strikes across the border, or taking the war into Germany. It meant ceding to Germany the initiative. France tragically made a similar choice in 1939-1940, when many experts agree that had France invaded Germany first, it might well have succeeded, or at least France would not have fallen as rapidly as it did.

President Harry Truman abstained from using nuclear weapons in Korea and from mobilizing fully the nation behind the war effort, which limited the military power the United States could bring to bear. He also ignored General Douglas MacArthur's calls to expand the war to China. Subsequent presidents similarly declined to mobilize the nation for the war in Vietnam, and they voluntarily limited the scope of American attacks on North Vietnam, Cambodia, and Laos. For much of the war, for example, the Haiphong port was off limits, even though mining and levelling the port, which was well within the United States military's capabilities, clearly would have constrained significantly North Vietnam's ability to sustain its war effort.

Israel's Golda Meir likewise refrained from any pre-emptive strikes against Egypt and Syria just before the 1973 war; like 1914 France, Meir reckoned that not appearing to be the aggressor was more important than whatever military advantage a 1967-style pre-emptive offensive might have achieved.

Decades later, Presidents George W. Bush and Barak Obama had the means to target Taliban fighters and leaders in their refuges in Pakistan. They also could have pressured Pakistan's leadership to cease helping the Taliban or hiding Taliban and Al Qaeda leaders. They thought it more important to preserve their bi-lateral relationship with Pakistan, even if that meant going along with the lie that Pakistan was not working hand-in-hand with the Taliban and Al-Qaeda. The most Obama was willing to do was conduct drone strikes in Pakistan, which at that point were too little and too late, and kill Bin Laden, to the Pakistani government's consternation.

In the current Israeli and American war against Iran, the two allies have chosen to focus on air power rather than risk major land operations. Of course, it would be very difficult for Israel to move against Iran on the ground, given geography, but the United States, thanks to its massive navy and its air lift capabilities, could. American restraint makes sense for a large number of reasons, but the fact of the matter is that the U.S. has decided the war does not justify applying the full weight of American military power. It has categorized the war as a limited one, just like all the wars the United States has fought since World War II.

The point is not to decry all of these choices as mistakes. Each was and is defendable. Nonetheless, they represent cases in which one side accepted limitations on its Liberty of Action. In at least some of these instances, the adversary acted to encourage or strengthen limitations, perhaps through propaganda or appeals to public opinion. Think of the outcry, for example, prompted by

President Richard Nixon's invasion of Cambodia in 1970, including the infamous protest at Kent State. Although it is difficult if not impossible to establish a causal relationship between North Vietnam's efforts to promote its narrative of the Cambodia invasion and the fact that a large number of people in the United States and abroad more or less shared the Communist narrative (people can draw their own conclusions), one reasonably can assume the North Vietnamese information operations were not irrelevant, either.

Initiative

Initiative arguably is something French strategists value above all else. It must be taken and retained; one must at all times seek to deny it to one's adversary. If one takes the initiative, one's adversary must react. In reacting, the enemy is doing something other than what it had planned to do. One therefore is limiting the enemy's Liberty of Action.

The criticality of the initiative goes a long way to explain French strategists' emphasis on the offensive, an emphasis that arguably went too far at the turn of the 20th century, the period when French Army doctrine could be summed up as *offensive à outrance*, or extreme offensive. Castex perhaps explained it best when he wrote that the offensive "represents action and movement." The offensive, he continued:

> ...transforms the power relationship. It modifies situations. It changes a stage of things to another that it seeks to realize. It engenders the novelty it conceives. It obliges birth. The offensive is, *par excellence*, a creative act.[4]

In contrast, for Castex the defensive could "only be static." At best, it prevented the enemy from "succeeding in its creative act." Castex even described the defensive as "an act of sterilization vis-à-vis the germs of life that tend toward the evolution of a crisis; it is an

---

4 Raoul Castex, *Théories Stratégiques*, IV (Economica, 1997), 91.

effort of non-transformation." [5] Or, to put it more bluntly, "the offensive imposes, the defensive suffers." [6] Castex of course, like Clausewitz and others before him, recognized that sometimes one had to go on the defensive. Among other things, Economy of Means obliges one to be on the defensive somewhere. The aim, however, always had to be to seize the initiative at the earliest opportunity, which invariably meant returning to the offensive. Similarly, the point of Economy of Means or even dispersing one's forces always had to be enabling offensive action somewhere else. Thus, even when one had no choice but to act defensively, or perhaps to attend to a Servitude that meant deviating from an offensive goal, one always had to have in mind the imperative to act in as offensive and "positive" a way that one "reasonably can conceive." [7]

History is replete with examples of successfully seizing and keeping the initiative, or of failing to do so, often with tragic results. Franco-British passivity in the face of the German threat in the 1930s up to the invasion of France is a stark example of the latter. The two allies abandoned their Liberty of Action and allowed Hitler to attack at a time and place and in a manner of his choosing. Japan famously took the initiative across the Pacific and maintained it until Midway and Guadalcanal (1942), at which point the United States seized the initiative and prevented the Japanese from regaining it ever again. Japan's Liberty of Action collapsed; its attempts to maneuver to increase it failed.

In contrast, Israel's pre-emptive strikes against Egypt and Syria in 1967 demonstrate dramatically the value of making the opening

---

[5] Castex, *Théories Stratégiques*, IV, 91.

[6] Castex, *Théories Stratégiques*, IV, 92.

[7] Raoul Castex, *Théories Stratégiques*, VI (Economica, 1997), 215.

14

moves to oblige the enemy to react rather than exercise its own will. Egypt and Syria turned the tables in 1973, with nearly disastrous results for Israel, until the Israelis rallied, counter-attacked, and regained the initiative. One can only presume that during the entire period in which Israeli forces were on the defensive, Israeli commanders never lost sight of the imperative of counterattacking at the earliest possible moment. Israel's beeper campaign in 2024 against Hezbollah in many ways was analogous to the pre-emptive air-strikes of 1967, in the sense that it was part of a planned strategy for seizing the initiative and forcing Hezbollah to react rather than set in motion its own plans. After all, even simply preventing an adversary from doing what it wanted, and forcing it to do something else, or nothing at all, is of immense strategic value.

A correlate to the imperative of seizing and maintaining the initiative is the need to act fast and keep acting fast. Among other things, this means being willing and able to make quick decisions. Decide, and decide fast. This implies having to trust one's gut and act notwithstanding insufficient information. As Foch repeated, "of all faults one alone is infamous, inaction."[8]

Once one has selected a course of action and executed, one has to follow up, quickly, and maintain as high an operational tempo as possible. Foch's relentless hammering of the Germans in 1918 after the failure of the German offensive is an example on an epic scale. At no point after the Allied offensive began on 18 July did the German Army manage to organize a major counterattack. In 1940, the French allowed the Germans to seize the initiative thanks to their overly defensive doctrine; once the Germans achieved that, the French were unable to recover in large part because the Wehrmacht's command cycle, according to which commanders received

---

[8] Maréchal Ferdinand Foch, *Des principes de la guerre* (Economica, 2007), 245.

information, decided, and communicates instructions to lower echelons, was radically faster than that of the French.

In 1973, the armies of Egypt and Syria opened their war against Israel with a surprise attack that gave them the initiative, to stunning effect: They successfully penetrated Israeli defenses on the Suez Canal and the Golan Heights. They in fact did much better than they had expected, which led to a fatal problem: The Syrians in particular had no orders instructing them as to what to do next, and both they and the Egyptians lacked the kind of command culture that encouraged field commanders to identify and act upon opportunities that might have enabled them to maintain the initiative. They paused. The Israeli Defense Forces, in contrast, were schooled in the need for speed and had a command culture that stressed agility and improvisation. They lost no time in rallying, counterattacking, and thus controlling the initiative until the war's end.

The American military's AirLand Battle doctrine of the 1980s emphasized moving fast and sustaining a high operational tempo to seize and retain the initiative in the face of superior enemy numbers. The first blow should achieve an element of surprise, perhaps by coming at an unexpected location, that would throw the adversary off-balance. Critically, one had to keep the enemy from recovering its balance by following up on the first blow with rapid and incessant moves. In the words of the 1986 Army Field Manual 100-5, "these operations must be rapid, unpredictable, violent, and disorienting." "The pace must be fast enough to prevent [the enemy] from taking effective counteraction." The United States deployed this doctrine to good effect in the 1991 war against Iraq.

French Admiral Guy Labouérie in 1992 wrote about the idea of surprise in similar terms, specifically using the words "uncertainty" and *foudroyance*. Uncertainty quite simply is something one most go to great lengths to cultivate among one's adversaries: uncertainty about what one is doing and going to do, where, when, and why. *Foudroyance*, derived from the word for thunder (*foudre*), means a sudden crippling shock:

> The principle of *foudroyance* has as its goal not destroying everything, which is without interest in any conflict, but breaking the rhythm or rhythms of the Other in its diverse activities, in such a way as to keep it from pulling itself together and to keep it a step behind the action.[9]

To do that, one must strike at the right moment, at just the right place, where the effect would be to block the enemy's attempt to retake the advantage or restore cohesion.

More recently, the French campaign in Mali in 2013 was a textbook example of using speed to seize and maintain the initiative. It was AirLand Battle in miniature. No one expected the French to intervene at all: The first elements of the French Army, heliborne, arrived as if out of the blue as a *deus ex machina* to stop advancing jihadist columns. Within a matter of days, the French switched from defensive operations to an aggressive offensive in which they used all of the mobility at their disposal—including airborne operations— to advance at breathtaking speed, pushing elements forward as they arrived in theater. While it is hard to predict what kind of resistance the Malian jihadists might have been able to give the French Army given the chance to regain their balance and perhaps take the initiative, the point is that the French never gave them that chance.

A major element of American thinking about AirLand Battle, which all Western militaries have embraced, is the idea of using technology to enable fast decision making. Often the decision-

---

[9] Vice-amiral d'escadre Guy Labouérie, *Stratégie: Réflexions et variations* (Addim, 1992), 179.

making cycle is discussed in terms of the so-called OODA Loop (Observe, Orient, Decide, and Act), a concept minted by U.S. Air Force officer John Boyd in the 1970s. You see something—perhaps a threat, or an opportunity, you figure out what it is you're seeing, you decide what to do about it, and then you do it. The Americans, concerned by their numerical weakness relative to the Soviets and the Warsaw Pact, gambled that they could use their technological superiority to speed up the OODA loop. They could do this through advanced sensors, information networks to collect and disseminated information, and precision weapons intended to achieve what un-guided weapons could do only a lot faster and with much less mass. Speeding up the OODA loop through technology has been a central obsession of Western militaries ever since. Interestingly, French military thinkers have expressed concern that technology might have the opposite effect, by encouraging commanders, habituated to an unprecedented wealth of information, to hesitate in the face of incomplete information and wait for still more data. At some point commanders have to make a call, meaning they have to trust their instincts. War, they stress, remains an art, an act of human creativity. It cannot be left to computers to use algorithms to calculate the best course of action based on data that in any event will never be complete.

Maneuver

The U.S. Army's definition of Maneuver often is stated as "the employment of forces through offensive or defensive operations to achieve relative positional advantage over an enemy force to achieve tactical, operational, or strategic objectives." This makes sense, but maneuver should be understood to mean a lot more than the physical movement of forces and "a relative positional advantage."

Castex again takes the lead here. He defined maneuver broadly as "as moving intelligently to create a favorable situation." He was less interested in a "relative positional advantage," however, than he was in restoring the initiative, and thereby increasing one's own Liberty of Action at the expense of the adversary's. The idea is to take the initiative to "modify or determine the course of events, to dominate destiny and not to abandon oneself to it, to engender and give birth to facts."[10] Indeed, Castex wrote, "one does not conduct a maneuver by being subject to the will of the enemy and by accepting the law of luck."[11] Maneuver, in Castex's eyes, was a "creative act" that modifies or determines the course of events."[12] Critically, he insisted that maneuver does not necessarily mean physical movement. It could just be an intellectual shift, a different way of thinking about problems.[13]

---

[10] Raoul Castex, *Théories Stratégiques*, II (Economica, 1997), 5.

[11] Castex, *Théories Stratégiques*, II, 5.

[12] Castex, *Théories Stratégiques*, VI, 192.

[13] Castex, *Théories Stratégiques*, VI, 192.

Beaufre carried the idea further by stressing the potency of diplomatic or political maneuvers, which, like military movements, could also improve one's situation by impinging upon the adversary's Liberty of Action and strengthening one's own. Beaufre wrote about the "artichoke maneuver," which amounted to the strategy of conducting small acts of aggression calculated not to trigger a major response, but which had the cumulative effect of greatly strengthening one's own position to the point where the prize falls into one's hand.[14] The example he most explicitly had in mind was Hitler's seizure of the Rhineland, Austria, and Czechoslovakia. Beaufre noted that military action in fact played only a small role in these acts of aggression: Hitler maneuvered diplomatically and psychologically, to ensure the paralysis of the British and French and render the Czechs isolated and impotent. Then, and only then, did he deploy force. Another concept Beaufre wrote about was the "lassitude maneuver," which amounted to building one's strategy around tiring out one's adversary.[15]

Most importantly, Beaufre described what he called the "external maneuver," discussed above. The external maneuver could often succeed at tying one's opponent's hands, thereby altering the balance of power on the battlefield. Maneuvers of any kind had the capacity to create possibilities. Having options amounted to having Liberty of Action. One could choose rather than being denied any choice.

Maneuver in this sense could be a question of figuring out a way to reduce an adversary's strengths and play on their weaknesses. That might mean going around rather than straight at an enemy force. But it could also be something as banal as the North Vietnamese government's invitation to Jane Fonda to visit North Vietnam to

---

[14] Beaufre, *Introduction à la stratégie*, 154.

[15] Beaufre, *Introduction à la stratégie*, 155.

20

"bear witness" and thus serve as a high-profile mouthpiece for North Vietnamese propaganda. Or managing who presents what information to the international press. Or attempting to have one's adversaries brought before the International Court of Justice. If one cannot stop one's adversaries from using advanced aircraft to attack one's positions, one perhaps can hamper their air operations by campaigning for sanctions that would deny them access to parts or munitions. Thus, in the Gaza war, Al Jazeera arguably was Hamas's most effective weapon.

In this light, the Israeli and American initiative known as the Abraham Accords represents a form of maneuver. The intention is to change the paradigm of relations between Israel and the Arab world in effect by marginalizing the Palestinian national movement. This is intended to diminish material and political support for Palestinian groups and increase Israel's own Liberty of Action. Whether it works remains to be seen.

The *Effet Majeur*

Closely related to the idea of striking a decisive point is the French concept of the *effet majeur*, which translates as "major effect." The term itself is misleading, for "major" implies aiming to achieve something big, perhaps like going after the heart of an enemy's defenses or its *Schwerpunkt*, to borrow from Clausewitz, the place where one's main force and effort should converge to do the most damage. What the French have in mind is something different.

Most of the definitions of the *effet majeur* that one finds in French military publications discuss it in terms of doing the thing that guarantees victory. For more precision, we turn to the recently retired General Michel Yakovleff, whose *Tactique Théorique (Theoretical Tactics)* is a standard reference among contemporary French officers. Yakovleff defines *effet majeur* in terms of doing the thing that allows one to seize a part of the initiative or break that of one's adversary.[16] The idea is that once one seizes the initiative, a number of things will result that will yield victory. Retired Colonel Michel Goya, who is now among the most prolific French military writers, offers even more precision. In Goya's words, the *effet majeur* is "the minimum one can do to be sure of fulfilling the mission."[17] Or, the minimum thing one can do that restores the initiative, leading then to a succession of steps that result in victory.

---

[16] General Michel Yakovleff, *Tactique Théorique*, 3rd ed. (Economica, 2016), 166, 175.

[17] https://x.com/Michel_Goya/status/1881382625283387520?s=20

French writers, including Yakovleff, compare this approach to what they see as the American understanding of the *Schwerpunkt*. They see this concept as encouraging one to attack directly whatever one has identified as the enemy's center of gravity. The problem with that is that the enemy's center of gravity often is its strongest point.[18] That could work if one has the means of attempting a direct strategy. Otherwise, one requires an Indirect Strategy, which requires in turn identifying that thing one *can* do that will set in motion a chain of events that will end with success. Yakovleff explains that when identifying how to achieve the *effet majeur*, one usually has to think past first-order effects and look instead to second- or third order ones. For example, imagine if one attacked a unit and destroyed it. The destruction of the unit is the first-order effect. That is a good thing, but what matters is not the destruction of the unit but the fact that by destroying the unit, one may have disrupted the enemy's plans. One has taken the initiative and denied it to the enemy. One has limited its Liberty of Action and follow up with actions that dictate what happens. This is how one wins.

Yakovleff cites the example of the German offensive in 1940 through the Ardennes. The Wehrmacht's advance through Sedan had the effect of destroying some French units. But that is not what mattered, especially given that the bulk of the French army, to the north and west, was intact. What mattered was that the German maneuver disrupted French plans, and from that point onward the French were never able to regain the initiative. That is what doomed the French army. To put it another way, Dislocating the plan is more important that destroying the combat power of any specific units. The destruction of units might be the necessary means to the end, but it is not the end, which is ultimately about initiative.

---

[18] Yakovleff, *Tactique Théorique*, 177–78.

Yakovleff also emphasizes that the *effet majeur* has to do with an action against the enemy, but not the terrain.[19] The point is never to seize a particular piece of ground, but to interfere with the enemy's plan in some meaningful way. That might require seizing a physical location, but the key is to think about the maneuver in terms of the enemy's intentions and how to interfere with them.

---

[19] Yakovleff, *Tactique Théorique*, 176.

The Primacy of Will

Foch, like his peers, identified the root cause of France's defeat in 1870 as a spiritual failing that translated into passivity and the lack of will to fight. Citing the conservative Catholic philosopher Joseph de Maistre, Foch wrote, "A lost battle is a battle one believes one has lost, for […] a battle is not materially lost."[20] For Foch, the opposite was also true: "A battle won is a battle in which one does not admit defeat." Thus, citing Foch, Beaufre defined strategy as the "art of the dialectic of wills that employ force to resolve their conflict."[21]

Indeed, like any good student of Foch, Beaufre placed great emphasis on the importance of will, of acting upon it, and of sapping that of the adversary. It follows that everything that has a psychological effect should be brought to bear, while military action, which usually is a material action taken upon material things, matters only in so far as it affects psychology. All the French strategists in fact make clear that destroying the enemy's armed forces is not necessary; what mattered is inducing them to believe the battle or the war is lost and give up.

The question then became identifying what it might take to do that. This is a question all the French strategists take up. On a tactical level, they emphasis the importance of surprise, which has the potential to induce paralysis. There is a moment when one's adversary may at least perceive that it has been stripped of its Liberty of Action. It cannot maneuver to improve its situation. There is

---

[20] Foch, *Des principes de la guerre*, 247.

[21] Beaufre, *Introduction à la stratégie*, 34.

nothing to be done. On a more strategic level, the task is one of convincing an adversary that further armed action is futile or can do no more to achieve a desired outcome. The only alternative, then, is to try to call it quits and turn to the negotiating table. One also has to think critically about the adversaries' weaknesses or vulnerabilities. Where are its true pain points? This line of thinking invariably brings these strategists to the matter of the relationship between war and politics. If war is politics by other means, to paraphrase Clausewitz, then the purpose of war necessarily is to achieve a political end state.

A critical part of the art of war, analytically speaking, is identifying how precisely military action can contribute to the achievement of a desired political result. France's post–World War II thinkers all agree that military action certainly can make a valuable and perhaps critical contribution but is unlikely on its own to achieve a "decision." There really is only so much military action can do. Thus, leaders have to identify what it can do presumably in tandem with multiple other lines of effort.

Spirit versus Matter

The French emphasis on will encourages skepticism regarding the importance of material means and even technology. It is not a question of what kind of arms one brings to a fight or how many, but of the relative strength of the opposing combatants' will to fight.

The realities of modern warfare, especially of the First World War, caused French military thinkers to moderate this view. The most notable example is Foch, who already by late 1914 conceded that the great strength of Germany's superior artillery negated the ability of French infantry's great morale to enable them to prevail on the battlefield. The more the war bogged down, the more Foch recognized that to a large degree winning battles had become a question of massed fires, which also was a question of industrial capacity. France needed enormous quantities of artillery, and gigantic quantities of shells for its artillery to fire. It needed this to neutralize German artillery so that French infantry could carry the day.

Foch never, however, renounced that idea that without greater will and an aggressive spirit, victory would not be possible. His own iron will arguably enabled him to cope with a succession of crises and rally wavering colleagues to grit their teeth and carry on. Castex wrote about Foch in the preface to a book published in 1943 that Foch "dreamed only of the initiative, the attack, without hindering himself with the prejudicial question of means." Castex continued:

> [Foch] propagated around him a surprising dynamism. He did not limit himself to resisting in place adverse fortune; he anticipated it to master it, to change its course, to replace it with another, *to create events instead of being subject to them*... He was—it is the greatest elegy one can give him—the doctor one calls in all the cases that are serious, if not hopeless, not only because of his professional value, but also because of his morale, which was inaccessible to discouragement, and his animating mentality. He, invested with such a frightening charge, only conceived of his function in the sense of an offensive rebound, toward which he gave all his force. [22]

In more recent decades, one can find echoes of this approach in French reticence regarding military technology. French observers of the American military often criticized Americans of being in the thrall of technology and assuming that the capabilities it delivered

---

[22] Raoul Castex, "Préface historico-militaire," in *Joffre, Foch, Gallieni*, by Louis Vetty (Éditions Chantal, 1943), 4. Emphasis mine.

would ensure victory. Yet, as the French point out, time after time since the Second World War, America's mighty military, for all its technological savvy, failed to overcome far less well-equipped opponents. One reason was that, in their eyes, Americans failed to understand the limits of military action in pursuit of political aims. Thus, decision in battle failed to result in strategic decision. Another reason was that the Americans failed to appreciate the strength of the adversaries' will. So impressed were the Americans with their unparalleled ability to crush the enemy tactically, that they did not give due consideration to what they really needed to be thinking about: How to defeat the enemy's will.

This was one of French theorists' biggest takeaways from the colonial wars of Indochina and Algeria, the idea that *all warfare was psychological warfare*, and all military actions needed to be viewed through the lens of their psychological effect. As Beaufre wrote,

> The key element of action is, first of all, the will of he who undertakes it. Then it is the will of those who could oppose that action. The psychological domain therefore is capital.[23]

This, Beaufre insists, is "where human factors" and "ideologies" intervene, and "material means" in and of themselves only serve to exploit these ideologies. He went so far as to announce that now the "psychological domain" was an "essential and decisive zone of action." Moreover, Beaufre cautioned that Western militaries are too materialistic in the philosophical sense. As a result, "we are more apt to calculate in tons of primary materials, dollars, means, troops, numbers of tanks and megatons, than in ideas, propaganda, and slogans."

---

[23] Andre Beaufre, *La Stratégie de l'action*, Kobo Ebook (A. Colin, 1966).

Indirect Strategy

French thinking on the concept of Indirect Strategy pre-dates the British strategist Basil Henry Liddell Hart's writings on the topic, but the French, to begin with Beaufre, who corresponded extensively with Hart, openly borrowed heavily from him.

Hart, haunted by the First World War and what he considered Britain's obscenely wasteful and fruitless ground operations in Belgium and France, often described the Indirect Strategy in tactical terms: Rather than attacking head-on the enemy's prepared positions, go around. Attack the flanks. Go behind it. Target logistical lines. Identify soft spots. Hart also emphasized trying to disrupt the enemy's equilibrium, perhaps through surprise, and playing on it psychologically.

Beaufre had a more expansive view of Indirect Strategy, one in which military action was only one of several instruments of possible maneuver, the others including diplomatic and economic actions. He saw this at work in the 1930s, watching Hitler use diverse means to strengthen Germany's hand and weaken that of the Allies prior to 1939, all without triggering a full war. It was then that Beaufre came up with the concept of "peace-war," the idea being that rather than seeing "war" and "peace" as distinct periods, one should imagine rival countries existing in a permanent state of "peace-war" in which each maneuvers through various means to enhance its Liberty of Action and weaken that of the other, often without resorting to armed conflict, or at least ensuring that acts of violence remain below a threshold liable to trigger open conflict.[24]

Beaufre's understanding of the "peace-war" that prevailed between France and Germany in the 1930s served him well intellectually when nuclear weapons seemingly changed the whole paradigm of military strategy. He saw nuclear weapons as making Indirect Strategy a necessity, because they prevented Great Powers from going to war directly against one another. Rather than pursue war in a "major key," war had to be conducted in a "minor key."[25] That meant, for example, acts of aggression that stayed below the threshold of triggering a nuclear response. This usually entailed attacking interests that were peripheral to an adversary, rather than vital. Militarily weak nations, moreover, learned to favor indirect approaches when battling superior nations. This was the case with the Viet Minh and the Viet Cong, and countless other "liberation" movements. Rather than seek decision in battles, which they were almost always certain to lose because of their military inferiority, they focused on surviving and tiring their opponents out. They sought to prevail in the contest of wills. They became adept at identifying ways outside of the theater of combat to sap their opponents' will, perhaps by targeting public opinion in the home country. They used propaganda and lawfare. They used international forums to disseminate their narratives and castigate their opponent, all things that had the effect, among other things, of limiting the enemy's Liberty of Action. Beaufre referred to this as the "external maneuver."[26] Against this, countries needed to deploy a counter-

---

[24] André Beaufre, "La Paix-Guerre ou la stratégie de Hitler," in *Écrits de combat 1939-1942* (Éditions Perrin, 2025). This was originally published in 1939.

[25] Beaufre, *Introduction à la stratégie*, 143.

[26] Beaufre, *Introduction à la stratégie*, 150.

external maneuver, as often it was the battle between external maneuver and counter-external maneuver that would decide the outcome of a war, not any physical battle in the theater of combat.

One is, of course, reminded of Vietnam, in which the United States won battle after battle, but to no avail. The American war effort was undone by policy choices that, among other things, resulted in self-imposed limitations on the military's Liberty of Action, and a divided public opinion that had proved vulnerable to Vietnamese Communist external maneuvers. The Communists' "lassitude maneuver," i.e. working to survive while tiring out the adversary, prevailed. Afghanistan was no different, although in that case it is hard to point to a Taliban "external maneuver." America's will dissolved all by itself in the face of the adversary's military strategy that amounted to playing for time.

Part of Beaufre's argument for Indirect Strategy has to do with his view that modern conflicts seldom if ever were an affair of two and only two warring countries. Many others were or might be a party to a conflict, and the two antagonists needed to appeal to the others to keep get them on side or at least induce them to stay out of the fight. Their success of failure in this endeavor was likely to be more decisive than anything that transpired on the battlefield.

All Wars Are Four-Way Wars

Beaufre imagined that in each conflict there were four camps.[27] There was one's own country, which Beaufre designated A. Allies with coinciding interests were B. One's principal adversaries were C. The rest of the world was E. E was not directly involved, but its comportment vis-à-vis the conflict between A and C could strongly

---

[27] Beaufre, *La Stratégie de l'action.*

influence the outcome of the conflict for good or for ill. E in fact can be decisive, either by joining A or C, or by opting to let the two sides fight it out. Basically, to win the war, A has to convince C of the futility of further conflict, with the support of B and all the while trying to enlist E or keep E from supporting C. A has to enlarge B, which means courting E. C, however, will have the same thing in mind, resulting in a tug of war between A and C over E.

If A were clearly superior to C, a direct strategy would be appropriate. Regarding E, all A had to do was keep E out of the conflict. But even then, the conflict could not end with a stable result without E's acquiescence, if not support. Beaufre stressed that usually it was in A's clear interest to act as rapidly as possible so as to be able, quickly, to present the world with a *fait accompli*. Beaufre saw Hitler's annexation of Czechoslovakia as a classic example of how this was done.

If A and C were evenly matched, the contest would require indirect strategies to enlist as much of E as possible into joining B. Eventually, once the balance of power clearly had become favorable, one switched to a direct strategy to decide the conflict. (Beaufre understood that this essentially was how the First World War played out.)

If A was clearly weaker than C, but believed it had the will to sustain a long fight, it would combine a direct action calculated to make the conflict last, while using Indirect Strategy to get E to act in ways favorable to it. In this case, it was the action on E that decided the war, i.e. the Indirect Strategy.

In the post-1945 era, Western powers needed American support for just about all of their military activities, even if that came in the form of Washington allowing its allies to carry on. France could not have pursued its wars in Indochina and Algeria without American material support, which often was begrudging given America's preference that France focus on its NATO obligations. It is for this reason, for example, that de Lattre in 1951 travelled to the United States to drum up support for the Indochina War. Not only did he meet with American leaders, but he also conducted public events and conducted numerous interviews with the American press.

Suez of course is perhaps the ultimate example of the price of not attending to Allies' interests. Militarily speaking, the Franco-British-Israeli campaign against Egypt was highly effective and successful. But the British and French had failed to get President Dwight Eisenhower's support, and Eisenhower in effect canceled the war. As Beaufre explained in his memoire of the war, Britain and France had failed to attend to this decisive diplomatic requirement (Castex would have called it a Servitude) and lay the groundwork for action. He believed Hitler in the 1930s had shown precisely how it needed to be done: First take the necessary steps to gain American support and isolate Egypt; then, act fast, so fast as to present the world with a *fait accompli*. Britain and France did none of these things.

More recently, the Falklands War provides a surprising example of how in the modern world (and perhaps well before that), even a seemingly straight-forward conflict between two antagonists involved outside powers and obliged the two (A and C, in Beaufre's parlance), to attend to B and E, either to keep their support or encourage them to stay out of the fight.

Britain needed American support (which was not necessarily a given, in light of warm relations between Washington and the Argentine junta, which it saw as a staunch anti-Communist ally, or past examples of American impatience with British imperial activities such as Suez). Sure enough, Reagan, perhaps more because of his friendship with Margaret Thatcher than anything else, provided ample help in the form of vital intelligence and logistical support in the form of aviation fuel, signals intelligence, and full access to the joint Anglo-American base on Ascension Island. Even if one could argue that the United Kingdom might have been able to pull off its victory without the Americans, they absolutely could not have done it had the United States opposed the war.

To a lesser yet important extent, Britain also needed Chilean and French help. Chile provided radar information vital to warning the British of impending Argentine air operations, as well as diplomatic support in Latin America. Perhaps most importantly, Chile pinned down significant Argentine military assets by maintaining a threatening posture. It is for this reason, for example, that Argentina did not deploy its best land forces to the islands, preferring to keep them in reserve to counter the Chilean threat. All Chile had to do to provide Argentina with vital assistance, if it wanted, was to change its military posture and work to reassure Buenos Aires.

France supported the British war effort by providing access to naval facilities in West Africa, sharing technical information regarding the Exocet antiship missiles it had sold to Argentina, and, perhaps most importantly, working to ensure that Argentine could not acquire any more.

Argentina understandably tried to internationalize the conflict and obtain help from fellow Latin American countries or developing nations in the name of anti-imperialism. With the exception of Chile, of course, Latin American countries generally sided with Argentina. However, rhetoric aside, they did little in terms of substantive support. Peru lent Argentina 10 Mirage V jets with their associated armaments, but they arrived just days before the war's end and never saw any action. It goes without saying that they might have done more, and that collectively, Latin American countries easily could have supplied a decisive amount of assistance. They did not. One has to wonder what kind of quiet diplomatic pressure Washington, London, and Santiago might have applied, or whether Buenos Aires might have obtained more help had worked to affect some sort of External Maneuver or simply tried harder. Being out maneuvered diplomatically compounded the problem of being out maneuvered in the air or on the land and sea by the Royal Navy, Royal Air Force, and British Army.

Today, the Ukraine War illustrates clearly Beaufre's framework at work. Ukraine's President Vladamir Zelensky from the start has understood the critical importance of international help. Ukraine's survival depends on securing from E material support while also encouraging sympathetic countries to apply sanctions to Russia and stoke international opprobrium. "I need ammunition, not a ride," he declared soon after the start of Russia's major offensive in February 2022, and since then he tirelessly has been making the rounds to international capitals and international forums. Ukraine has had a hard time with President Donald Trump, but Zelensky clearly has understood the crucial importance of cultivating his relationship with him, as well as with prominent American lawmakers. Russia likewise has been using various means to undermine Western support for Ukraine, including through disinformation strategies and bot farms, while working to obtain material support from sympathetic neutrals or circumvent international sanctions. In other words, the war between Ukraine and Russia is taking place not only on the Ukrainian battlefield but in the court of world opinion and the halls of foreign parliaments. Both countries are actively engaged in a variety of External Maneuvers that complement their battlefield maneuvers, and the former may prove as decisive as the latter, especially given Ukraine's dependency on foreign-supplied weaponry and intelligence.

Likewise, the war in Gaza pitted both sides in fierce competition over E while obliging them to manage their relations with B. Israel obviously needed American support above all, but also needed continued support or at least inaction on the part of many countries and international organizations that were friendly or not-so-friendly, from Germany to Saudi Arabia. Israel needed them to preserve as much of its Liberty of Action as possible. To that end, Israel deployed diplomatic means and public diplomacy. For its part, Hamas has to attend to its backers, Qatar among them, while it and the many people sympathetic to its cause abroad participated in an extensive External Maneuver to isolate Israel and reduce its Liberty of Action.

The current Iran war also illustrates Beaufre's vision, although more in the guise of an example of what not to do. Washington, rather than cultivating allies, friends, and other third parties, has gone out of its way to alienate many of those it might have counted upon to lend a hand. The Trump Administration's hostility toward America's long-standing allies reflected a belief that the United States had little need of allies, who in any case had little to contribute. "American First" was tantamount to "America alone." Nonetheless, the war thus far has demonstrated American reliance on its allies and friends. First, it has needed at the very least the complicity of European allies and regional friends to stage its military operations. America has been flowing aircraft and supplies to the Middle East through European bases; it has been conducting many of its operations from European bases as well as bases located in regional friends like Jordan. In the meantime, the British have been hesitating, no doubt because of Trump's rough treatment of them, although they ultimately have been allowing the United States to make use of its bases in the United Kingdom. Those countries most capable of contributing to the campaign against Iran, France and Italy, likewise are hesitating. France has mobilized, but less to help the Americans and Israelis than to reassure Greece and its friends in the Gulf. Italy, other than contributing a frigate to the French fleet in the

Mediterranean, pointedly is staying out of the fight. Spain, too, could lend a hand given its military might, but other than supplying a frigate of its own to help the French, has concentrated on attacking Israel diplomatically. Under another president, the United States should have been able to organize and lead combined fleets involving multiple partners to operate in the Eastern Mediterranean and the Arabian Sea. Perhaps even to force the Straits of Hormuz. A French-led fleet is forming, but it most likely will be doing its own thing.

Does America need an allied fleet? It would help. Also, in one crucial way the U.S. Navy, operating without allied navies, is in a real fix: The Navy has very limited anti-mine capabilities, which might lock it out of the Straits. This is not an oversight: It has not invested in anti-mine capabilities specifically because the U.S. and NATO assigned that mission to European navies. It was seen as a niche capability for smaller countries that could not or should not be asked to acquire fleets large surface combatants for so-called blue water operations, because of their great cost.

Total Strategy

Castex and Beaufre both argued for the necessity of having a Total Strategy, of which a military strategy was only one component. Implied is the need to understand military strategy's place, i.e. its limits, or rather what military action could and could not do to advance policy. Castex appears to have derived his ideas chiefly from his readings of military theorists; for Beaufre, this was a lesson bitterly learned in France's campaigns in Indochina, Suez, and Algeria.

Castex opens his magnum opus *Théories Stratégiques,* with a metaphor that conveys his overall view of military strategy. "Strategy," he writes, "is like the light spectrum." He continues:

> It has an infrared, which is the realm of politics,
> and an ultraviolet, which is that of tactics. Just as
> the spectrum connects these invisible parts with
> undetectable gradations, strategy is joined to
> politics, and tactics progressively alter to blend
> with them. Politics, strategy, and tactics thus
> form a whole, complete, and in no way a triptych
> of clearly separated elements.[28]

Throughout *Théories* Castex stresses the interconnectivity of different strands of strategy and the need for all the strands to be understood as a whole. Jumping to the sixth volume of *Théories,*

---

[28] Raoul Castex, *Théories Stratégiques*, I (Economica, 1997), 11.

Castex evokes the "principle of the *unity of war*."[29] "War is *one*," he continues, "and it always has been this way." "War," moreover, "is *total*, as one likes to say these days." The emphasis here is Castex's. By "total," Castex does not mean the kind of warfare we saw in the First and Second World Wars in which adversaries mobilized all resources in an all-out struggle to destroy their adversaries to the point of leveling their cities and starving their populations. He is thinking more about the mobilization of the full range of resources, even if the total level of effort might be limited.

Castex explains that the "unity of war finds expression in the will to fight of a nation, from which all the rest flows, and which moves together the gears of the machine."[30] The gears are separate domains (economic, diplomatic, naval, etc.), and for each there are specific strategies. But above them all there must be a "general strategy," which is the strategy for conducting all the forces and means of a nation."[31] The general strategy "goes beyond, dominates, coordinates, and disciplines the particular strategies."[32] There must be some arbitrage among the different strategies, for sometimes one requires another to include an action that otherwise would not be a priority. This generates what he describes as the "servitudes," discussed below.

It is not clear if Beaufre ever read Castex, but he also wrote about the need for a total or Grand Strategy. "The only good strategy," he insisted in *Strategy of Action*, "is total." One element of this was a reappreciation of Clausewitz's insight that war was "politics by other means." Beaufre went to great lengths to emphasize that military

---

[29] Castex, *Théories Stratégiques*, VI, 1.

[30] Castex, *Théories Stratégiques*, VI, 2.

[31] Castex, *Théories Stratégiques*, VI, 2.

[32] Castex, *Théories Stratégiques*, VI, 2.

42

action always must be subordinate to politics and seen as but one part of a larger cluster of actions that a country could and should undertake to achieve desired political ends. This was, he argued in his memoir of the Suez War, one of the key lessons from that debacle, in which France, Britain, and Israel conspired to toppled Egyptian President Gamal Abdel Nasser and seize the Suez Canal without preparing the ground diplomatically and politically, thereby dooming the campaign regardless of anything that might be achieved in battle. Beaufre certainly would not endorse the American penchant for delegating strategy to the military (Vietnam, Iraq, Afghanistan), and repeatedly expecting military action to yield the desired political outcome. Total strategy meant subordinating military strategy to a comprehensive strategic concept, which itself is dictated by the political concept and elaborated and executed by politicians. In *Strategy for Tomorrow* he explained:

> Military war generally is no longer decisive in the literal sense of the word. Political decision, always necessary, can only be obtained through a combination of limited military action with appropriate actions taken in the psychological, economic, and diplomatic domains. The strategy of war, previously governed by military strategy, which for a while gave preeminence to military leaders, now depends on a total strategy led by the heads of the government, with military strategy only playing a subordinate role.[33]

Like any good student of Foch, Beaufre placed great emphasis on the importance of will, of acting upon it, and of sapping that of the adversary. It follows that everything that has a psychological effect should be brought to bear, while military action, which usually

---

[33] André Beaufre, *Strategy for Tomorrow* (Macdonald and Jane's, 1974), 3.

is a material action taken upon material things, matters only in so far as it affects psychology. And it is imperative that military action complement all the other actions taken in parallel or taken beforehand to prepare the military action.

Servitudes

The idea of the Servitudes did not originate with Raoul Castex, but it is he who elevated it to a major part of French military thinking, and it is he with which the concept is associated. If asked about Castex, most French officers today would link him to the Servitudes.

Castex defined Servitudes as "obligations foreign to the strategy of the milieu," milieu here meaning the strategy of a specific military branch.[34] Later, Castex revised this definition. "Servitudes for a determined strategy, are obligations that are foreign to its normal field of action, but that nonetheless merit serious consideration."[35]

The basic concept is that leaders could not give free rein to military strategy and conduct war in a vacuum. The military might say that militarily speaking, the best course of action would be to do X. The problem was that, in reality, one had to accommodate other imperatives. These could be political imperatives, diplomatic imperatives, legal imperatives, etc., all of which required deviating from the straight line of thinking informed purely by military considerations.

To give a crude example, think of all the conflicts in which the simplest and probably most effective strategy would be to kill everyone, yet one does not. Why not? The reasons why often are Servitudes, i.e. because one risks losing domestic or international support, etc. Or perhaps one could settle something quickly by deploying one's entire military, or by using nuclear weapons. Imagine

---

[34] Castex, *Théories Stratégiques*, I, 241.

[35] Castex, *Théories Stratégiques*, VI, 3.

if Prime Minister Thatcher had simply threatened Argentina, saying, "either evacuate your troops from the Falklands by noon today or we will nuke Buenos Aires." It would have worked, and thereby saved a lot of time, money, and British lives. The idea almost certainly never occurred to Thatcher, and she undoubtedly would have refused had someone suggested it, even if, practically speaking, that was the simplest, shortest, and most effective strategy. There were other considerations.

Likewise, the United States could have deployed its entire military to Afghanistan and perhaps have gotten better results. Presidents Bush, Obama, Trump and Biden, however, had good reasons for not doing that, among them the need to keep large portions of the military available for possible other tasks they regarded as more important. They accepted the necessity of going about the war in Afghanistan with limitations on the number of troops to send, troops required for other obligations.

Castex cited two major examples from the First World War. The first was the German decision to violate Belgian neutrality and invade France through Belgium. Militarily, this made sense: It was the shortest line between two points. Politically and diplomatically, however, the move virtually guaranteed Britain's entry into the war. As Castex described it, Germany's leaders heeded military strategy without attending to diplomatic and political Servitudes. Of course, the United Kingdom might have joined the war on France's side even if Germany had side-stepped Belgium, but the Germans did not *know* if it would. Castex believed that if the Germans could have done something, anything, to avoid the United Kingdom's entry into the war, that would have been worth trying, given the difference Britain's participation in the war made for the balance of power. The second example was Germany's decision to engage in unrestricted submarine warfare. As he described it, throughout the war there was a fight within the German government between a 'submarine lobby' that saw submarines as the single best tool for weakening Britain and France, especially given that the Royal Navy had bottled up Germany's surface fleet. If one wanted to defeat Britain and France, unrestricted submarine warfare offered the best chance of success. However, others within the government insisted on minding the diplomatic or political Servitude of not bringing the United States into the war. The submarine lobby eventually won out; Germany unleashed its U-Boats; the United States declared war, all but guaranteeing German defeat if the war lasted long enough for the Americans to bring their strength to bear. Which is precisely what happened.

In more concrete terms, the Servitudes might oblige a naval commander to refrain from massing all of its assets into a single fleet to go out and seek the enemy on the high seas, because the "morale" Servitude required calming public opinion by allocating some ships to defend the coastland. It might not make sense militarily, but it soothed the public, which could be of strategic value for the war effort. The example Castex cites is the American Navy's sense of obligation during the Spanish-American War to keep back some of its fleet to guard the coastline in response to the widespread fear that the Spanish fleet might attack. For Castex, this was necessary, even though in strictly military terms it made far more sense not to bother, given that the American Navy needed all its strength for a possible fleet action in the Atlantic. The principle of Concentration of Force argues strongly in favor of not dividing one's fleet.

Castex's concept of the Servitudes, however, reminds one that sometimes one should divide one's forces to cover a secondary theater or attend to some other requirement critical to the overall war effort. Or, perhaps rather than attack one area, which made the most military sense, one needs to attack another because of its economic value. That was an economic Servitude. Alternatively, there often were various reasons why one could not use any weapons or any means to achieve an objective. Massacring civilians, for example, could be militarily useful, but has a price in terms of public opinion or morale. In contemporary times, such considerations go a long way to explain the rarity of poison gas on the battlefield, or why the United States did not use nuclear weapons in Korea. Or why, today, Trump does not simply threaten to nuke Tehran. He could; it would work, if the Iranians thought he was serious. But even Trump understands why that would be a terrible idea.

Castex described Servitudes in the context of the simultaneous existence of separate strategies that all operate (hopefully in concert) as part of a total or grand strategy. There are land strategies and naval strategies, economic strategies, and diplomatic strategies, etc. These multiple strategies make demands of one another. "Servitudes," he wrote, "are the actions and reactions of diverse strategies." "They come," he continued, "from the unity of war, the existence side by side of a military war, a political war, an economic war, a war of morale, etc." He adds, "The Servitudes are the internal and necessary links of this ensemble that derive from the notion of total war."[36]

The challenge posed by the Servitudes is that danger of losing sight of one's primary objectives in war. "If one cedes to these servitudes without discernment and above all a necessary method…[strategy] easily forgets its own permanent goal."[37] Go ahead and attend to Servitudes, Castex wanted to say, but always keep in mind what they are and how they relate to strategy; do not lose sight of the bigger picture and what one wants to do with one's naval forces. The catch was that to do that, one must have a clear

---

[36] Castex, *Théories Stratégiques*, I, 251.

[37] Castex, *Théories Stratégiques*, I, 249.

idea of what one's objectives really are. One must have a strategy if one is going to decide whether attending to any particular Servitude was necessary.

More contemporary examples of countries attending to Servitudes or ignoring them at their peril abound. Diplomatic and political Servitudes kept the United States during the Korean and Vietnam War from availing itself of the full means at its disposal, like using nuclear weapons, full military mobilization, and or devastating their adversaries through conventional means. There would have been a heavy diplomatic price to pay, not to mention a significant cost in terms of public support. The British and French, as mentioned above, failed to attend to diplomatic Servitudes prior to launching their Suez campaign. Israel in its response to the 7 October attacks arguably acted in contempt of the many Servitudes that might have encouraged different strategies or tactics. For example, the Israeli military might have legitimate military justification for attacking a certain location, never mind the collateral damage or simply how its attack might play on the BBC. The reputational damage could prove decisive, if it undermined the support of Israel's allies and strengthens the External Maneuvers of Israel's adversaries, who seek to limit Israel's Liberty of Action and diminish Israel's military advantages. It appears to be the case that Israel's leadership, in seeking to enlarge its military's ability to maneuver on the battlefield, might have seen that ability shrink because of their failure to mind key Servitudes. Why did Israel do this? A likely answer appears to be Prime Minister Benjamin Netanyahu's concern for political Servitudes, namely keeping his political coalition together. Beaufre would instruct him to weigh if attending to these political Servitudes overly hinders military strategy, or if it truly was necessary for prosecuting the war and achieving its objectives.

Concluding Remarks

The above pages provide a clear vision of military strategy, one that emphasizes initiative, maneuver, and will. This is not an industrial strategy, nor a recipe for prevailing in wars of attrition. If anything, it calls for doing everything possible to avoid wars or attrition or to break the impasse once one has fallen into that trap. How? Not through force, but through creativity. By "maneuvering intelligently" to improve a situation. That could be direct or indirect; on the battlefield or by means of some form of external maneuver.

Strategy above all has the following four requirements:

1. A clear idea of the desired political end state
2. A clear idea of what kind of total strategy might bring about such an end state
3. A clear idea of the military's role within the larger total strategy: What precisely can military force accomplish in complementarity with all the other aspects of strategy?
4. Means matched with ends: If the means are inadequate to achieve a desired goal, the ends must be redefined.

And all throughout, Castex would have us mind the Servitudes. They do not have to be catered to. They have to be identified and weighed but never ignored.

The current war against Iran leaves me wishing the insights discussed in this small volume were shared by the civilian and military decision makers at the helm of the American and Israeli war efforts. Both countries' leadership have failed to articulate desired political end states. Both have failed to consider what kind of total strategy required to bring about a desired political end state, which is impossible if one cannot articulate one. (Beaufre: "There can be no strategy other than total strategy").[38] Both have failed to define how precisely military action can support that total strategy. Indeed, they seem to rely entirely on military action, a form of direct strategy Beaufre goes to great lengths to explain cannot work.

Both, moreover, appear to have neglected the factor of will, i.e. what actions one might take to strengthen one's own side's will and undermine that of the adversary. History is replete with examples proving that aerial bombardment and material destruction are insufficient. Nor have they considered that strategy, per Beaufre, is the "art of the dialectic of wills." Iran could be expected to attempt its own maneuvers, and to employ whatever means at its disposal to undermine its adversaries' will and expand or at least protect its Liberty of Action. Thus far everything the Iranian government has done, from attacking its neighbors' oil infrastructure to terrorizing civilian populations with missile and drone attacks to closing the Strait of Hormuz were obvious and predictable. Yet somehow the Americans in particular appear to have been surprised.

Finally, both Jerusalem and Washington seem uninterested in identifying let alone addressing the many Servitudes that beg for attention in this conflict. These include economic Servitudes related to the war's enormous impact on the global economy, diplomatic Servitudes related to minding relations with friends and allies, and

---

[38] Beaufre, *Introduction à la stratégie*, 25.

political Servitudes tied to building and maintaining public support. Both countries' leaders appear to have decided they could charge ahead like Admiral David "Damn the torpedoes" Farragut in Mobile Bay, confident that everything would go not as planned—for they had not planned—but as they hoped. One need not have read Beaufre or Castex to recognize that hope is not a strategy.

Bibliography

Beaufre, André. "La Paix-Guerre ou la stratégie de Hitler." In *Écrits de combat 1939-1942*. Éditions Perrin, 2025.

Beaufre, Andre. *La Stratégie de l'action*. Kobo Ebook. A. Colin, 1966.

Beaufre, André. *Strategy for Tomorrow*. Macdonald and Jane's, 1974.

Beaufre, General André. *Introduction à la stratégie*. Pluriel, 2012.

Castex, Raoul. "Préface historico-militaire." In *Joffre, Foch, Gallieni*, by Louis Vetty. Éditions Chantal, 1943.

Castex, Raoul. *Théories Stratégiques*. I. Economica, 1997.

Castex, Raoul. *Théories Stratégiques*. II. Economica, 1997.

Castex, Raoul. *Théories Stratégiques*. IV. Economica, 1997.

Castex, Raoul. *Théories Stratégiques*. VI. Economica, 1997.

Foch, Maréchal Ferdinand. *Des principes de la guerre*. Economica, 2007.

Géré, François. *La pensée stratégique française contemporaine*. Economica, 2017.

Labouérie, Vice-amiral d'escadre Guy. *Stratégie: Réflexions et variations*. Addim, 1992.

Pierre, General Hervé. *Le Général Beaufre: Père de la stratégie française*. Perrin, 2025.

Yakovleff, General Michel. *Tactique Théorique*. 3rd ed. Economica, 2016.